Real-time Analytics in Managed Pressure Drilling

ISBN: 9798895877500

Table of Contents

Chapter 1: Introduction to Managed Pressure Drilling

Overview of Managed Pressure Drilling

Managed Pressure Drilling (MPD) is an innovative drilling technique that enables operators to precisely control the annular pressure profile throughout the drilling process. This method offers a significant advancement over traditional drilling practices by allowing real-time adjustments to the pressure in the wellbore. By maintaining optimal pressure levels, MPD minimizes the risks associated with wellbore instability, kicks, and lost circulation. The technique is particularly valuable in challenging geological formations, where conventional drilling may lead to costly delays and safety hazards.

Real-time data analytics plays a crucial role in the implementation of MPD. By integrating real-time data collection with advanced analytical tools, operators can monitor and react to downhole conditions instantaneously. This capability not only enhances decision-making but also improves operational efficiency. Data from various sources, including surface measurements, downhole sensors, and geological models, are analyzed to provide insights that guide drilling parameters, such as weight on bit, mud properties, and flow rates. Such analytics help in optimizing drilling performance and reducing non-productive time.

The environmental impact and sustainability of drilling operations are increasingly important considerations for the oil and gas industry. MPD contributes to these goals by reducing the environmental footprint of drilling activities. By minimizing the risks of fluid loss and enhancing well control, MPD leads to lower chances of surface spills and groundwater contamination. Furthermore, the ability to drill more efficiently can reduce the overall energy consumption associated with drilling operations, thereby contributing to more sustainable practices within the industry

Advanced equipment and technologies are at the core of effective MPD operations. The integration of automated systems, real-time monitoring tools, and advanced drilling rigs enhances the capability to manage pressure dynamically. Equipment such as automatic chokes, pressure sensors, and

advanced mud systems play a pivotal role in achieving the desired pressure profiles. The continuous evolution of technology in this domain enables the oil and gas industry to tackle increasingly complex drilling challenges while ensuring safety and compliance with regulatory standards.

Case studies from successful MPD projects highlight the effectiveness of this drilling method in diverse environments. These projects demonstrate how MPD has been successfully employed in offshore drilling, high-pressure high-temperature wells, and areas with complex geological features. Analyzing these real-world applications offers valuable lessons on best practices, risk management, and the integration of technology in MPD. Additionally, understanding regulatory compliance and safety standards in managed pressure drilling is essential for researchers and practitioners to navigate the legal landscape and ensure the safe execution of drilling operations.

Importance in the Oil and Gas Industry

The oil and gas industry operates in an environment characterized by complexity and high stakes, making the importance of real-time analytics in managed pressure drilling increasingly vital. This advanced methodology allows operators to make informed decisions based on accurate, up-to-the-minute data, significantly reducing the risks associated with drilling operations. By leveraging real-time data analytics, companies can optimize drilling performance, enhance safety measures, and minimize environmental impacts, all of which are crucial for maintaining operational efficiency and regulatory compliance.

Real-time analytics provide a framework for continuous monitoring of various parameters during the drilling process.

This includes pressure, temperature, and fluid properties, which are critical for managing wellbore stability and preventing hazardous situations such as blowouts. By integrating sophisticated data models and predictive analytics, operators can anticipate potential issues before they escalate, thus ensuring the safety of personnel and equipment. The ability to react swiftly to changing conditions not only safeguards investments but also fosters a culture of proactive risk management within the organization.

The role of real-time analytics extends beyond immediate operational safety; it also contributes to sustainable practices within the oil and gas sector. The environmental impact of drilling activities has garnered increasing scrutiny, making it essential for companies to adopt technologies that mitigate risks to ecosystems. Real-time data analytics facilitate the identification of environmentally sensitive areas and enable the implementation of strategies to reduce emissions and waste. As such, the integration of these analytics can lead to more environmentally responsible drilling practices, aligning with global sustainability goals.

Moreover, the competitive landscape of the oil and gas industry demands that companies optimize their drilling operations continually. Real-time analytics empower professionals to identify inefficiencies in drilling techniques and equipment performance, leading to cost savings and improved resource allocation. By analyzing historical and current data, companies can refine their drilling strategies, ensuring that they remain competitive in an ever-evolving market. This adaptability is particularly crucial as energy demands fluctuate and regulatory environments become more stringent.

Finally, the importance of real-time analytics in managed pressure drilling is underscored by the need for compliance with safety standards and regulations. As regulatory bodies impose stricter requirements on drilling operations, the integration of real-time data analytics becomes essential for demonstrating compliance. By systematically collecting and analyzing data, companies can provide evidence of adherence to safety protocols and environmental regulations. This not only mitigates legal risks but also enhances the overall reputation of the organization in an industry where public perception is increasingly tied to environmental and safety performance.

Key Concepts and Terminology

Managed Pressure Drilling (MPD) represents a transformative approach in the oil and gas industry, particularly in enhancing drilling efficiency and safety. Key concepts in MPD include the control of the annular pressure profile, which allows for the maintenance of a constant bottom hole pressure. This concept is crucial for preventing well control issues and optimizing drilling performance. The ability to dynamically adjust the mud properties and pressure levels in real time distinguishes MPD from traditional drilling methods, enabling operators to respond swiftly to subsurface conditions. Understanding these foundational elements is vital for researchers and professionals aiming to master advanced techniques in this field.

Real-time data analytics is another essential concept in MPD, involving the continuous monitoring and analysis of data collected during the drilling process. This encompasses a variety of data streams, including pressure, volume, and temperature measurements, as well as geological and operational parameters. By leveraging advanced algorithms and machine learning techniques, operators can extract meaningful insights from this data, enhancing decision-making capabilities and improving operational outcomes. Familiarity with various data analytics tools and methodologies is imperative for those in academia and industry, as these technologies are integral to the effective implementation of MPD strategies.

The terminology associated with MPD includes several specialized terms that are crucial for communication among stakeholders. Terms such as "constant bottom hole pressure," "choke management," and "rate of penetration" are frequently used in discussions surrounding drilling operations. Understanding these terms allows professionals to effectively engage in technical discussions, contribute to research initiatives, and collaborate on innovative projects. Additionally, comprehension of regulatory compliance terminologies, such as "environmental impact assessments" and "safety standards," is essential for ensuring that MPD practices align with industry regulations and sustainability goals.

In the context of environmental impact and sustainability, key concepts include the assessment of ecological risks associated with drilling activities, as well as the implementation of best practices to mitigate these impacts. Sustainable MPD practices involve not only optimizing resource extraction but also minimizing the environmental footprint of drilling operations. Researchers and industry professionals must be well-versed in sustainability metrics and frameworks to effectively evaluate and enhance the environmental performance of MPD projects.

This knowledge is increasingly demanded by regulatory bodies and stakeholders who prioritize ecological responsibility alongside operational efficiency.

Finally, case studies on successful MPD projects provide valuable insights into the practical application of these key concepts and terminologies. Analyzing real-world examples allows researchers and practitioners to identify best practices, challenges encountered, and solutions implemented in various drilling environments. Such case studies often highlight the interplay between advanced equipment, real-time data analytics, and regulatory compliance, demonstrating how these elements converge to achieve successful outcomes. By studying these instances, professionals can gain a deeper understanding of the complexities involved in managed pressure drilling and the critical importance of terminology and concepts in driving innovation within the industry.

Chapter 2: Mastering Managed Pressure Drilling

Advanced Techniques in Drilling

Advanced techniques in drilling have evolved significantly with the integration of real-time analytics, which enhances decision-making and operational efficiency in managed pressure drilling (MPD). These innovative methods include advanced mud management systems, real-time monitoring of downhole conditions, and the use of adaptive drilling technologies. Each of these elements plays a crucial role in optimizing drilling performance, reducing non-productive time, and improving safety. By leveraging real-time data, operators can make informed adjustments to drilling parameters, ensuring that operations remain within the desired pressure and flow limits.

One notable advancement is the implementation of intelligent drilling systems that utilize machine learning algorithms to analyze vast amounts of data generated during the drilling process. These systems can predict potential issues, such as kicks or blowouts, by identifying patterns and anomalies in the data. This predictive capability allows for proactive measures to be taken, reducing risks and enhancing overall safety. Moreover, these systems are designed to continuously learn from each drilling operation, refining their predictive models to improve future performance

Another critical aspect of advanced drilling techniques is the development of high-performance drilling fluids tailored for specific geological conditions. These fluids are engineered to maintain optimal downhole pressure while providing effective cooling and lubrication to the drill bit. The formulation of these fluids often includes environmentally friendly components to minimize the ecological impact of drilling activities. By employing advanced fluid management strategies, operators can enhance the stability of the wellbore and reduce the risk of formation damage, ultimately leading to more efficient drilling operations.

Furthermore, the use of remote sensing technologies and real-time data transmission has transformed the operational landscape of managed pressure drilling. These technologies facilitate instant communication between the drilling rig and remote monitoring centers, allowing for real-time analysis of drilling parameters. By integrating satellite and IoT-based solutions, operators can track critical indicators, such as pressure, temperature, and flow rates, in real time. This immediate access to data enables rapid response to any deviations from expected performance, thus ensuring adherence to safety standards and regulatory compliance

•

Finally, case studies of successful managed pressure drilling projects illustrate the practical application of these advanced techniques. These projects often highlight the synergy between technology and human expertise, showcasing how a well-trained workforce equipped with cutting-edge tools can achieve remarkable results. By analyzing these case studies, researchers and professionals can gain insights into best practices and innovative approaches that can be adopted in their own operations. The continuous evolution of advanced drilling techniques, supported by real-time analytics, not only enhances operational efficiency but also contributes to the sustainability goals of the oil and gas industry.

Technologies Transforming Drilling Practices

Technological advancements are reshaping drilling practices in the oil and gas industry, particularly in the context of managed pressure drilling (MPD). Innovations in real-time data analytics, for instance, have drastically improved decision-making processes during drilling operations. By utilizing sophisticated algorithms and machine learning techniques, operators can analyze vast amounts of data from various sources instantaneously. This capability allows for the detection of anomalies and real-time adjustments to drilling parameters, enhancing efficiency and reducing the risks associated with conventional drilling methods.

Another significant advancement is the integration of automated systems into drilling rigs. Automation not only optimizes drilling performance but also minimizes human error, which is crucial in high-pressure environments. Automated control systems are now capable of adjusting mud weight, flow rates, and other operational parameters dynamically, based on real-time data. This level of automation not only improves operational consistency but also enhances safety by allowing human operators to focus on strategic decision-making rather than routine tasks.

The development of advanced drilling tools and equipment has further transformed drilling practices. Managed pressure drilling employs specialized equipment such as rotating control devices, dual-gradient systems, and advanced blowout preventers. These technologies facilitate more precise control over downhole conditions, enabling drilling operations to navigate challenging geological formations with greater ease. As a result, companies can achieve optimum drilling performance while minimizing environmental impacts, aligning with industry goals of sustainability and regulatory compliance.

Moreover, the shift towards cloud computing and the Internet of Things (IoT) has revolutionized data management in drilling operations. By leveraging cloud platforms, companies can store and analyze drilling data more efficiently and collaboratively. IoT sensors integrated into drilling equipment provide continuous monitoring and data collection, enabling operators to make informed decisions based on real-time conditions. This connectivity not only enhances operational transparency but also supports predictive maintenance strategies, reducing downtime and operational costs.

Lastly, advancements in simulation and modeling technologies are playing a crucial role in pre-drilling planning and risk assessment. Sophisticated modeling software can simulate various drilling scenarios, allowing engineers to anticipate potential challenges and devise effective strategies before actual drilling begins. This proactive approach minimizes the environmental impact and ensures compliance with safety standards, ultimately resulting in more successful drilling projects. As these technologies continue to evolve, they will further enhance the efficacy of managed pressure drilling practices, solidifying their importance in the oil and gas industry.

Challenges and Solutions in Implementation

The implementation of real-time analytics in managed pressure drilling (MPD) presents a myriad of challenges that professionals in the oil and gas industry must navigate. One significant hurdle is the integration of diverse data sources into a cohesive analytics platform. The complexity arises from the variety of sensors, equipment, and technologies used in drilling operations, which often produce data in different formats and at various frequencies. This fragmentation can lead to difficulties in data synchronization and interpretation, resulting in potential inefficiencies and compromised decision-making processes. Addressing this challenge necessitates the development of robust data integration frameworks that can seamlessly aggregate and standardize data from multiple sources.

Another critical challenge lies in the need for real-time processing capabilities. The dynamic nature of drilling operations requires that data analytics be performed instantaneously to enable timely responses to changing conditions. However, many existing systems struggle to deliver real-time insights due to limitations in computational power and data transmission speeds. To overcome this barrier, the adoption of edge computing technologies can be beneficial.

By processing data closer to the source, edge computing reduces latency and enhances the speed at which critical information is delivered to decision-makers on the ground.

Moreover, the implementation of real-time analytics systems must consider the reliability and accuracy of the data being analyzed. In the high-stakes environment of MPD, erroneous data can lead to catastrophic failures or safety incidents. Thus, establishing rigorous data validation and quality assurance protocols is paramount. These protocols should include regular audits of data sources, real-time monitoring of sensor performance, and the use of advanced algorithms that can identify and correct anomalies in the data stream. By ensuring data integrity, professionals can make informed decisions that enhance operational safety and efficiency.

Training and skill development represent another challenge in the successful implementation of real-time analytics. The workforce in the oil and gas sector often requires upskilling to effectively utilize advanced analytics tools and interpret the insights generated. A comprehensive training program that focuses on both the technical aspects of data analytics and the specific applications within MPD is essential. Collaborations between industry stakeholders and academic institutions can facilitate knowledge transfer and equip professionals with the necessary skills to leverage real-time analytics effectively.

Finally, regulatory compliance and safety standards must be integrated into the implementation process of real-time analytics. As the industry evolves, so too do the regulations governing drilling operations. Ensuring that analytics systems comply with these regulations while also promoting safety requires an ongoing dialogue between technology developers and regulatory bodies. By proactively engaging with regulators and incorporating compliance considerations into the design and deployment of analytics solutions, companies can foster a culture of safety and responsibility that aligns with industry best practices.

Chapter 3: Real-time Data Analytics in Managed Pressure Drilling

Importance of Real-time Data

The importance of real-time data in managed pressure drilling (MPD) cannot be overstated, as it serves as the backbone for decision-making processes in complex drilling environments. Real-time data enables practitioners to monitor well conditions continuously, facilitating immediate responses to unexpected changes in pressure, temperature, and fluid dynamics. This capability is crucial in mitigating risks associated with well control incidents, which can lead to costly delays and environmental hazards. By harnessing real-time data, drilling teams can maintain optimal pressure levels and enhance the overall efficiency of operations, thereby ensuring the safety and sustainability of drilling activities.

Real-time data analytics in MPD allows for a more profound understanding of subsurface conditions and reservoir behavior. By integrating data from various sensors and systems, professionals can develop predictive models that inform drilling strategies. This analytical approach not only helps in identifying potential issues before they escalate but also aids in optimizing resource allocation and reducing non-productive time. The ability to visualize and interpret real-time data enhances collaboration among team members, fostering a culture of proactive problem-solving and innovation in drilling techniques.

Additionally, the integration of advanced technologies such as machine learning and artificial intelligence in real-time data analytics significantly improves the accuracy of predictions related to drilling performance. These technologies can analyze large datasets from previous projects to identify patterns and correlations that human analysts may overlook. By applying such insights in real-time, drillers can adjust their strategies dynamically, leading to improved drilling rates and reduced operational costs. Moreover, this technological synergy supports continuous learning and adaptation, which are essential in the rapidly evolving landscape of the oil and gas industry

•

The environmental impact and sustainability of managed pressure drilling are also greatly enhanced by the use of real-time data. With the ability to monitor emissions and waste generation continuously, operators can implement measures to minimize their ecological footprint. Real-time insights facilitate compliance with environmental regulations and standards, ensuring that drilling operations align with sustainability goals. Furthermore, by optimizing drilling parameters through data analysis, companies can reduce energy consumption and resource depletion, contributing to more environmentally responsible practices within the industry.

Finally, case studies on successful MPD projects underscore the significance of real-time data in achieving regulatory compliance and safety standards. Projects that have effectively utilized real-time data demonstrate how proactive monitoring and management can lead to safer drilling environments and improved regulatory outcomes. These examples serve as valuable lessons for researchers and professionals aiming to refine their methodologies and technologies. By prioritizing real-time data analytics, stakeholders in the oil and gas industry can enhance their operational efficiency, uphold safety protocols, and support sustainable development initiatives, ultimately transforming the future of managed pressure drilling.

Data Acquisition Techniques

Data acquisition techniques play a pivotal role in managed pressure drilling (MPD) by facilitating the real-time monitoring and analysis essential for optimizing drilling operations. In this context, data acquisition encompasses a variety of methodologies and technologies designed to collect, transmit, and store data from multiple sources within the drilling environment. The effectiveness of these techniques directly influences decision-making processes, risk management, and overall operational efficiency, making it imperative for researchers and professionals to understand the different approaches available.

One of the most prevalent techniques in data acquisition is the use of sensors and instrumentation deployed at various points along the drilling assembly. These devices measure critical parameters such as pressure, temperature, flow rates, and mud density, providing continuous streams of data that are crucial for real-time analytics. Advances in sensor technology have led to improved accuracy and reliability, allowing for better detection of anomalies and enhancing the capabilities of predictive analytics models. Integration of these sensors with telemetry systems ensures that data is transmitted in real time to surface control centers, where it can be analyzed and acted upon promptly.

Another essential technique is the application of advanced data logging systems that compile and archive vast amounts of historical and real-time data. These systems not only facilitate the immediate analysis of drilling conditions but also contribute to long-term data repositories that can be leveraged for future projects. By employing machine learning algorithms and big data analytics, researchers can extract valuable insights from this historical data, identifying trends and patterns that can inform future drilling strategies and enhance operational safety. This retrospective analysis is particularly beneficial in understanding the environmental impact of MPD operations, as it allows for the evaluation of past practices and their effects on surrounding ecosystems.

Moreover, the utilization of cloud-based platforms for data acquisition has revolutionized the way data is managed and analyzed in the oil and gas industry. These platforms provide scalable storage solutions and enable real-time collaboration among teams, regardless of their geographical locations. By harnessing the power of cloud computing, professionals can access and analyze large datasets from multiple drilling operations simultaneously. This capability enhances the sharing of best practices and promotes a culture of continuous improvement within the industry, which is crucial for achieving regulatory compliance and maintaining safety standards.

Lastly, the integration of Internet of Things (IoT) technology into data acquisition systems has opened new frontiers in managed pressure drilling. IoT devices facilitate the connection of various equipment and sensors, creating a comprehensive network that enables seamless data sharing and communication. This interconnectedness enhances situational awareness and allows for the implementation of automated responses to specific drilling conditions, thereby reducing human error and improving safety outcomes. As the industry continues to evolve, staying abreast of these innovative data acquisition techniques will be essential for researchers and professionals dedicated to advancing the field of managed pressure drilling.

Analytical Tools and Software

Analytical tools and software play a crucial role in enhancing the efficiency and effectiveness of managed pressure drilling

(MPD) operations. These technologies allow researchers and professionals in the oil and gas industry to collect, process, and analyze vast amounts of real-time data from drilling operations. By employing sophisticated algorithms and data analytics techniques, these tools can help identify patterns, predict outcomes, and optimize drilling parameters. Consequently, the integration of advanced analytical tools not only supports decision-making processes but also contributes to minimizing operational risks and ensuring compliance with safety and environmental regulations.

The landscape of analytical software used in MPD is diverse, ranging from data visualization platforms to predictive modeling applications. Data visualization tools enable engineers and geoscientists to present complex drilling data in an easily interpretable format, facilitating better communication among team members and stakeholders. Meanwhile, predictive modeling software utilizes historical data and machine learning techniques to forecast future drilling performance, enabling proactive adjustments to drilling parameters. This predictive capability is particularly valuable in managing the inherent uncertainties associated with subsurface conditions, thereby enhancing drilling efficiency and reducing non-productive time.

Moreover, the implementation of cloud-based analytical solutions has revolutionized data management in MPD. Cloud technology allows for the centralized storage and real-time sharing of data across various stakeholders, including field personnel, engineers, and management teams. This accessibility ensures that all parties have up-to-date information, which is essential for making timely decisions in fast-paced drilling environments. Additionally, cloud-based platforms often incorporate advanced security features to protect sensitive data, ensuring compliance with industry regulations while supporting collaborative efforts across geographically dispersed teams.

Integration of artificial intelligence (AI) and machine learning (ML) into analytical tools further enhances their capabilities in the context of MPD. These technologies can automate the analysis of drilling data, identifying anomalies and inefficiencies that may not be apparent through traditional methods. By leveraging AI and ML, professionals can develop more accurate models of well behavior and optimize drilling strategies in real-time. This not only improves operational performance but also contributes to sustainability efforts by minimizing waste and reducing the environmental impact associated with drilling activities.

As the oil and gas industry continues to evolve, staying abreast of the latest analytical tools and software is essential for researchers, professionals, and scientists involved in managed pressure drilling. Investment in these advanced technologies not only enhances drilling performance and safety but also plays a vital role in ensuring regulatory compliance and sustainability in operations. Continuous education and training in the use of these tools will empower industry stakeholders to harness the full potential of real-time data analytics, ultimately driving innovation and efficiency in managed pressure drilling practices.

Chapter 4: Environmental Impact and Sustainability

Assessing Environmental Footprint

Assessing the environmental footprint of managed pressure drilling involves a comprehensive evaluation of the various impacts associated with the entire drilling process, from site preparation to well abandonment. This assessment is critical for researchers and professionals in the oil and gas industry, as it provides insights into the ecological consequences of drilling activities and informs strategies for minimizing adverse effects. The evaluation process encompasses a range of factors, including emissions of greenhouse gases, potential contamination of groundwater, and the disruption of local ecosystems. By employing real-time data analytics, industry stakeholders can monitor these factors continuously, enabling proactive management of environmental impacts.

One of the primary components of assessing the environmental footprint is the analysis of emissions generated during drilling operations. Managed pressure drilling (MPD) techniques, while designed to optimize efficiency and safety, also produce emissions that must be quantified and monitored. This includes evaluating the release of volatile organic compounds (VOCs) and methane, both of which contribute to air pollution and climate change. Utilizing advanced sensor technologies and data analytics, operators can gain real-time insights into emission levels, allowing for timely adjustments and the implementation of mitigation strategies.

Water management is another critical element in assessing the environmental footprint of managed pressure drilling. The potential for groundwater contamination is a significant concern, particularly in areas where drilling occurs near aquifers or sensitive water resources. Assessments must include an analysis of the methods used for water sourcing, use, and disposal, as well as the effectiveness of containment measures to prevent spills and leaks. Real-time monitoring of water quality can provide essential data to identify contamination events early, enabling swift remedial actions and compliance with regulatory standards.

Biodiversity impacts are also a crucial consideration in the environmental assessment process. The disruption of habitats due to drilling activities can lead to significant ecological consequences, particularly in sensitive areas. An effective assessment should involve evaluating the potential impacts on local flora and fauna, including endangered species. By leveraging real-time data analytics, operators can track changes in wildlife behavior and habitat use, allowing for adaptive management practices that minimize disruption and promote ecological resilience

Finally, the regulatory framework governing managed pressure drilling necessitates a thorough understanding of compliance standards related to environmental impact. Researchers and professionals must stay informed about local and international regulations that dictate environmental performance requirements. By integrating real-time analytics into compliance monitoring, organizations can ensure adherence to environmental laws and standards while fostering a culture of sustainability within the industry. This proactive approach not only enhances operational efficiency but also contributes to the overall goal of reducing the environmental footprint associated with drilling activities.

Sustainable Practices in Drilling

Sustainable practices in drilling have gained increasing importance in response to growing environmental concerns and regulatory pressures within the oil and gas industry. Managed Pressure Drilling (MPD) techniques have been at the forefront of this evolution, enabling operators to minimize their environmental footprint while maintaining operational efficiency. The integration of real-time data analytics plays a critical role in identifying and implementing sustainable practices, ensuring that drilling operations are not only effective but also environmentally responsible.

One of the key strategies in promoting sustainability within MPD is the optimization of drilling parameters through real-time data analysis. By continuously monitoring variables such as pressure, flow rate, and temperature, operators can adjust their drilling practices to reduce waste and energy consumption. For instance, real-time analytics can help in identifying the optimal mud weight and drilling fluid properties, which can lead to less fluid loss and reduced need for disposal. This proactive approach not only conserves resources but also mitigates the potential for environmental contamination.

Furthermore, advancements in equipment technology have paved the way for more sustainable drilling operations. Innovations such as automated drilling systems and advanced blowout preventers contribute to enhanced safety and efficiency. These technologies facilitate better control over drilling operations, reducing the likelihood of spills and other environmental hazards. Additionally, the use of eco-friendly drilling fluids and materials is becoming more prevalent, helping to minimize ecological disruption and promote a more sustainable approach to resource extraction.

Collaboration among industry stakeholders is essential for fostering sustainable practices in drilling. Research institutions, regulatory bodies, and oil and gas companies must work together to develop guidelines and best practices that prioritize environmental stewardship. Case studies showcasing successful implementations of sustainable MPD practices can serve as valuable learning tools, demonstrating the feasibility and benefits of adopting environmentally conscious approaches within the industry.

In conclusion, sustainable practices in drilling are essential for aligning the oil and gas industry with contemporary environmental standards and expectations. By leveraging real-time data analytics, advanced technologies, and collaborative efforts, the industry can achieve a balance between operational efficiency and environmental responsibility. As research continues to evolve in this domain, a commitment to sustainability will be pivotal in shaping the future of managed pressure drilling and the broader energy sector.

Mitigation Strategies and Innovations

Mitigation strategies and innovations in managed pressure drilling (MPD) are crucial for enhancing operational efficiency while minimizing environmental impact. As the oil and gas industry faces increasing scrutiny regarding its environmental footprint, the integration of real-time analytics has emerged as a pivotal tool for identifying and addressing potential risks during drilling operations. By leveraging advanced data analytics, operators can predict and manage downhole conditions more effectively, leading to reduced non-productive time and enhanced safety. This proactive approach not only aids in compliance with regulatory standards but also fosters a culture of sustainability in drilling practices.

One of the key innovations in MPD is the development of sophisticated monitoring systems that utilize real-time data to optimize pressure management. These systems are equipped with sensors that continuously collect data on various parameters, including formation pressure, flow rates, and fluid properties. This data is analyzed using advanced algorithms that can detect anomalies and provide actionable insights. By enabling operators to make informed decisions on-the-fly, these innovations significantly reduce the likelihood of blowouts and other hazardous situations, thereby improving operational safety and efficiency.

Furthermore, the application of machine learning and artificial intelligence in MPD represents a significant leap forward in predictive analytics. These technologies can analyze historical drilling data alongside real-time inputs to forecast potential challenges, such as stuck pipe incidents or kicks. By anticipating these issues, operators can implement targeted mitigation strategies that enhance drilling performance. For instance, adjusting the mud properties or modifying drilling parameters in real-time can lead to smoother operations and lower costs. This level of foresight is particularly beneficial in complex geological formations where traditional drilling techniques may fall short.

In addition to technological innovations, collaboration among industry stakeholders is essential for the development of effective mitigation strategies. Sharing best practices and case studies from successful MPD projects can provide valuable insights into overcoming common challenges. Industry forums and workshops focusing on real-time analytics can facilitate knowledge exchange, allowing professionals and researchers to explore new approaches and refine existing methodologies. This collaborative environment not only accelerates the adoption of innovative practices but also fosters a shared commitment to safety and sustainability in the oil and gas sector.

Lastly, regulatory compliance plays a critical role in shaping the mitigation strategies employed in MPD. As regulations become increasingly stringent, operators must adopt comprehensive risk management frameworks that incorporate real-time data analytics. This alignment with safety standards not only enhances operational integrity but also builds public trust in the oil and gas industry. By prioritizing innovation and sustainability through informed decision-making, the industry can navigate the complexities of modern drilling while ensuring environmental stewardship and operational excellence.

Chapter 5: Advanced Equipment and Technologies

Overview of Essential Equipment

The successful implementation of managed pressure drilling (MPD) relies heavily on the integration of essential equipment designed to facilitate real-time data acquisition, monitoring, and control. At the core of MPD operations are specialized drilling rigs equipped with advanced technology that allows for precise management of downhole conditions. This includes adjustable surface back pressure systems, which are crucial for maintaining wellbore stability and preventing blowouts. These systems enhance the ability to manage reservoir pressures effectively, enabling operators to drill more efficiently and safely.

Another critical component of MPD is the use of real-time data analytics tools, which provide continuous monitoring of various parameters during drilling operations. These tools are equipped with sophisticated sensors and data acquisition systems that gather information on pressure, temperature, and flow rates. The integration of these data analytics platforms facilitates immediate decision-making, allowing professionals to adjust drilling parameters on-the-fly. This capability not only enhances operational efficiency but also minimizes environmental risks associated with traditional drilling methods.

In addition to real-time analytics, the role of advanced downhole tools cannot be overlooked. Tools such as managed pressure drilling control systems, measurement while drilling (MWD) devices, and logging while drilling (LWD) equipment contribute significantly to the overall success of MPD. These technologies provide critical insights into the geological formations encountered during drilling, enabling operators to adapt their strategies accordingly. Furthermore, advancements in telemetry systems have improved the speed and reliability of data transmission from downhole tools to surface systems, ensuring that information is available for timely analysis.

The environmental impact and sustainability of MPD operations are also influenced by the essential equipment employed. Innovations in equipment design focus on reducing waste and enhancing the efficiency of drilling fluids, which play a vital role in minimizing the ecological footprint of drilling activities. Biodegradable drilling fluids and closed-loop systems are examples of technologies that contribute to more sustainable practices. By utilizing such equipment, companies can adhere to regulatory compliance and safety standards, ensuring that their operations meet the increasing demands for environmental stewardship within the oil and gas industry.

Lastly, the selection and deployment of essential equipment in managed pressure drilling must consider the unique challenges presented by different geological environments. Case studies of successful MPD projects illustrate the importance of tailoring equipment choices to specific operational needs. The interplay between equipment capabilities and environmental conditions can significantly affect drilling outcomes. As the industry continues to evolve, ongoing research and development will be vital in enhancing the performance and efficiency of MPD systems, ultimately leading to safer and more sustainable drilling practices in the future.

Innovations in Drilling Technology

Innovations in drilling technology have transformed the landscape of managed pressure drilling (MPD), enabling more efficient and safer operations in the oil and gas industry. Recent advancements have focused on enhancing the precision and responsiveness of drilling techniques, allowing for real-time adjustments to drilling parameters. These innovations are not merely incremental improvements; they represent a paradigm shift that integrates cutting-edge technologies such as automation, artificial intelligence, and advanced materials into the drilling process. The convergence of these technologies facilitates optimized drilling performance, minimizes non-productive time, and enhances overall project efficiency.

One significant innovation is the integration of real-time data analytics into drilling operations. Through the use of advanced data acquisition systems, operators can monitor drilling parameters in real time, allowing for immediate adjustments to be made to control the downhole environment effectively. This capability is crucial in managing wellbore stability and mitigating risks associated with unexpected pressure fluctuations. The utilization of machine learning algorithms further enhances this process by identifying patterns and predicting potential issues before they arise, thereby reducing the likelihood of costly downtime and improving safety outcomes.

Another noteworthy advancement is the development of automated drilling systems. These systems utilize robotics and automation technologies to perform repetitive tasks, such as tripping operations and making connections, with high precision and speed. Automation not only increases operational efficiency but also reduces the risk of human error, which is a significant factor in drilling incidents. Furthermore, the implementation of automated systems allows for remote monitoring and control, enabling operators to oversee drilling operations from distant locations, thus enhancing safety and reducing operational costs.

The material sciences have also contributed to innovations in drilling technology, particularly through the development of advanced drilling fluids and composite materials for drill bits and casing. These materials exhibit superior performance under extreme conditions, such as high temperatures and pressures, which are common in deep-water and unconventional drilling scenarios. Enhanced drilling fluids can provide better lubrication and cooling, reducing friction and wear on drilling equipment. The utilization of these advanced materials not only improves drilling efficiency but also extends the lifespan of critical equipment, ultimately leading to more sustainable drilling practices.

As the industry moves forward, the emphasis on sustainability and environmental impact will continue to drive innovations in drilling technology. New methods are being explored to minimize the ecological footprint of drilling operations, such as reducing the volume of waste generated and improving water management practices. The incorporation of eco-friendly materials and processes into drilling operations is becoming increasingly important as regulatory standards evolve. By embracing these innovations, the oil and gas industry can achieve a balance between operational efficiency and environmental stewardship, ensuring that managed pressure drilling remains a viable and responsible method for resource extraction in the future.

Future Trends in Equipment Development

Future trends in equipment development for managed pressure drilling (MPD) are poised to significantly enhance operational efficiency, safety, and environmental sustainability within the oil and gas industry. As the demand for more efficient drilling techniques escalates, the integration of advanced technologies into drilling equipment is becoming increasingly critical. Innovations such as automated control systems, real-time data analytics, and enhanced sensor technologies are leading the charge in transforming traditional drilling approaches. These advancements not only improve the accuracy of pressure management but also contribute to the overall reduction of operational risks associated with drilling activities.

One key trend is the miniaturization and integration of sensors within drilling equipment. As equipment becomes more compact, the ability to install multiple sensors on a single unit allows for real-time data collection and analysis. This shift enables operators to monitor critical parameters, such as pressure, temperature, and flow rates, with unprecedented precision. The ability to analyze this data in real time can lead to quicker decision-making processes, enhancing drilling performance and mitigating potential hazards. Furthermore, the proliferation of Internet of Things (IoT) technologies is facilitating seamless communication between equipment, allowing for a more holistic view of the drilling environment.

Another significant trend is the development of predictive analytics models that leverage machine learning algorithms. These models can analyze historical drilling data to predict potential failures or inefficiencies in real time. By anticipating issues before they arise, operators can implement preventive measures, reducing downtime and minimizing the environmental impact of drilling operations. Additionally, as these predictive models become more sophisticated, they can be integrated into equipment design, resulting in machinery that is not only more resilient but also more responsive to changing operational conditions.

The push for sustainability within the oil and gas sector is also influencing equipment development in MPD. Manufacturers are increasingly focused on creating environmentally friendly technologies that minimize emissions and reduce the ecological footprint of drilling operations. Innovations such as closed-loop systems and advanced waste management technologies are being integrated into drilling rigs to enhance sustainability. As regulatory compliance becomes more stringent, the development of equipment that adheres to these standards is essential, ensuring that companies can operate without compromising environmental responsibilities.

Finally, collaborative research and development efforts between academia, industry professionals, and equipment manufacturers are likely to shape the future landscape of MPD technologies. By pooling resources and expertise, stakeholders can accelerate the innovation cycle, leading to more effective solutions tailored to the specific challenges faced in managed pressure drilling. The establishment of cross-disciplinary partnerships will not only enhance the technological capabilities of drilling equipment but also foster a culture of continuous improvement that is essential for the long-term success and sustainability of the oil and gas industry.

Chapter 6: Case Studies on Successful Managed Pressure Drilling Projects

Overview of Selected Projects

This section provides an overview of selected projects that exemplify the application of real-time analytics in managed pressure drilling (MPD) across various contexts in the oil and gas industry. These projects highlight the integration of advanced technologies and data-driven methodologies that enhance operational efficiency, safety, and environmental stewardship. By showcasing a range of initiatives, this overview aims to illustrate the diversity of applications and the potential for innovation within the field.

One notable project involved the implementation of real-time data analytics in a deepwater drilling operation in the Gulf of Mexico. By utilizing advanced sensors and data processing algorithms, the project team was able to monitor drilling parameters dynamically, allowing for immediate adjustments to the mud pressure. This proactive approach not only minimized the risk of kicks and blowouts but also optimized drilling speed, leading to significant cost savings. The successful application of real-time analytics in this project serves as a benchmark for future endeavors in similar challenging environments.

Another significant initiative took place in the North Sea, where a consortium of companies collaborated on a managed pressure drilling project aimed at enhancing environmental sustainability. The project employed real-time monitoring systems to assess the ecological impact of drilling operations continuously. By analyzing data related to emissions, waste management, and local biodiversity, the team was able to implement strategies that reduced the environmental footprint of their activities. This project underscores the importance of integrating sustainability considerations into the operational framework of managed pressure drilling.

In the onshore sector, a project in the Permian Basin demonstrated the efficacy of real-time analytics in improving operational safety and compliance with regulatory standards. The implementation of a comprehensive data analytics platform allowed the drilling team to track and analyze compliance metrics in real time. This capability not only facilitated adherence to safety regulations but also fostered a culture of accountability and continuous improvement within the team. The project highlights how real-time analytics can play a critical role in regulatory compliance and safety management.

Furthermore, a series of case studies from various international projects illustrate the versatility of real-time analytics in different geological and operational contexts. These case studies reveal how tailored analytics solutions can address specific challenges faced by drilling teams, such as wellbore stability, fluid dynamics, and equipment reliability. By documenting the outcomes and lessons learned from these projects, researchers and professionals can gain valuable insights into best practices and innovative approaches that can be applied to future managed pressure drilling operations.

In summary, the selected projects presented in this overview exemplify the transformative potential of real-time analytics in managed pressure drilling. By leveraging advanced technologies and data-driven insights, these initiatives not only enhance operational efficiency and safety but also contribute to environmental sustainability and regulatory compliance. As the industry continues to evolve, the lessons learned from these projects will be instrumental in shaping the future of managed pressure drilling practices worldwide.

Lessons Learned from Each Case

The analysis of various case studies within managed pressure drilling (MPD) has revealed a multitude of lessons that can significantly enhance the deployment of real-time analytics in this field. Each case presents unique challenges and solutions that contribute to the collective knowledge base. From the integration of advanced technologies to the importance of regulatory compliance, these lessons serve as invaluable insights for researchers and professionals aiming to refine their approaches in MPD.

One of the primary lessons gleaned from the case studies is the critical role of real-time data in decision-making processes. In instances where operators utilized advanced data analytics tools, there was a marked improvement in the ability to predict and mitigate downhole issues. This proactive approach not only minimized costly delays but also enhanced the overall safety of drilling operations. The ability to analyze real-time data allowed teams to respond swiftly to changing conditions, thereby reducing the risks associated with pressure fluctuations.

Another significant takeaway is the necessity of collaboration among multidisciplinary teams. Successful MPD projects frequently involved engineers, data scientists, and environmental specialists working closely together. This collaborative environment fostered innovative solutions that addressed both technical and environmental challenges. For instance, in one case, the integration of environmental impact assessments into the real-time analytics framework led to the development of more sustainable drilling practices, demonstrating that interdisciplinary cooperation can yield beneficial outcomes.

Moreover, the case studies underscored the importance of regulatory compliance and safety standards in the effective implementation of MPD. Companies that prioritized adherence to local and international regulations not only avoided legal repercussions but also built trust with stakeholders and communities. This commitment to safety and compliance was often reflected in enhanced operational efficiency. By establishing stringent safety protocols and ensuring ongoing training for personnel, organizations were able to create a culture of safety that permeated all levels of operation.

Additionally, the evaluation of advanced equipment and technologies revealed that investment in the latest innovations is essential for optimizing performance in managed pressure drilling. Case studies showcased how cutting-edge tools, such as automated drilling systems and advanced sensors, contributed to greater precision and control over drilling parameters. These technologies not only facilitated more efficient drilling operations but also provided valuable data that informed future projects. As the industry continues to evolve, staying abreast of technological advancements will be crucial for maintaining competitive advantage.

In conclusion, the lessons learned from each case study in managed pressure drilling highlight the multifaceted nature of this field. Emphasizing real-time data analytics, fostering interdisciplinary collaboration, adhering to safety standards, and investing in advanced technologies are all vital components for success. As researchers, professionals, and scientists delve deeper into the complexities of MPD, these insights will serve as a guiding framework for enhancing operational efficiency, ensuring environmental sustainability, and achieving regulatory compliance in future projects.

Best Practices for Future Projects

In the realm of managed pressure drilling (MPD), implementing best practices for future projects is paramount for enhancing efficiency, safety, and sustainability. As the industry increasingly shifts towards integrating real-time analytics, it is essential to establish a structured approach that leverages advanced techniques and technologies. This begins with a comprehensive understanding of the operational environment, which includes meticulous planning and evaluation of the geological conditions, equipment capabilities, and potential risks. Researchers and professionals must prioritize data-driven decision-making to optimize drilling performance while minimizing environmental impact.

Collaboration among multidisciplinary teams is a critical best practice for future MPD projects. Engaging experts from various fields—such as geoscience, engineering, data analytics, and environmental science—fosters an environment of innovation and comprehensive problem-solving. The integration of real-time data analytics into the decision-making process allows for immediate adjustments to drilling parameters, thereby enhancing operational efficiency. Additionally, fostering open communication channels among team members ensures that knowledge and insights are shared, leading to more informed strategies and improved project outcomes.

Investing in advanced equipment and technologies is also vital for the success of managed pressure drilling projects. Organizations should prioritize the acquisition of state-of-the-art tools that enhance real-time monitoring and automation capabilities. This includes advanced downhole sensors, data acquisition systems, and integrated software platforms that facilitate seamless data flow and analysis. By utilizing these technologies, professionals can better predict and respond to potential challenges during drilling operations, ultimately reducing downtime and operational costs while maintaining a strong focus on safety and regulatory compliance.

Another essential practice involves continuous training and development for personnel involved in MPD projects. As technology evolves and the complexities of drilling increase, ongoing education becomes critical. Implementing training programs that emphasize the importance of real-time data analytics, environmental stewardship, and safety protocols ensures that team members are well-equipped to handle the dynamic nature of managed pressure drilling. Furthermore, fostering a culture of learning and adaptability within organizations empowers professionals to embrace new methodologies and technologies as they emerge.

Lastly, conducting thorough post-project evaluations is crucial for refining practices and enhancing future performance. By analyzing the successes and challenges faced during a project, teams can identify areas for improvement and establish benchmarks for subsequent endeavors. Documenting lessons learned and case studies not only contributes to the body of knowledge within the field but also aids in developing best practices that can be shared with the broader community. This iterative approach to project evaluation and improvement is essential for driving innovation and ensuring the long-term sustainability of managed pressure drilling within the oil and gas industry.

Chapter 7: Regulatory Compliance and Safety Standards

Overview of Regulatory Framework

The regulatory framework governing managed pressure drilling (MPD) is a complex and multifaceted structure that aims to ensure safety, environmental protection, and operational efficiency within the oil and gas sector. This framework encompasses a variety of local, national, and international regulations that dictate the operational standards and practices of MPD. These regulations are designed to address the unique challenges posed by MPD, including the need for precise control over wellbore pressure, the management of drilling fluid properties, and the minimization of environmental impact. Understanding this framework is crucial for researchers, professionals, and scientists involved in the development and implementation of MPD technologies and practices.

At the national level, regulatory bodies such as the Bureau of Safety and Environmental Enforcement (BSEE) in the United States and the Oil and Gas Authority (OGA) in the United Kingdom play a pivotal role in establishing guidelines for drilling operations. These agencies impose strict requirements for risk assessment, safety management systems, and environmental stewardship. Compliance with these regulations is mandatory, and companies must demonstrate their ability to manage the risks associated with MPD effectively. This includes conducting thorough environmental impact assessments and implementing robust safety protocols to mitigate potential hazards.

Internationally, organizations such as the International Association of Drilling Contractors (IADC) and the American Petroleum Institute (API) provide additional layers of guidance through best practice manuals and recommended practices. These documents serve as valuable resources for industry stakeholders, offering insights into standardized procedures for MPD operations. The collaboration between regulatory bodies and industry organizations helps to harmonize standards across different regions, promoting the adoption of advanced technologies and practices that enhance safety and efficiency in MPD.

Environmental regulations are particularly significant in the context of MPD, given the industry's increasing focus on sustainability. Many countries have enacted legislation aimed at reducing greenhouse gas emissions and protecting natural resources, which directly influences drilling practices. Companies engaged in MPD must navigate these regulations carefully, ensuring that their operations not only comply with legal requirements but also align with broader sustainability goals. This necessitates the integration of real-time data analytics to monitor environmental parameters and optimize drilling processes, thereby minimizing the ecological footprint of operations.

In conclusion, the regulatory framework surrounding managed pressure drilling is essential for ensuring the safe and sustainable advancement of this technology within the oil and gas industry. Researchers and professionals must remain informed about evolving regulations and industry standards to foster innovation while maintaining compliance. By understanding the interplay between regulatory requirements and operational practices, stakeholders can contribute to the development of more efficient, safe, and environmentally responsible MPD solutions. This comprehensive understanding is vital for the continued success and acceptance of managed pressure drilling as a leading technique in the industry's pursuit of operational excellence.

Safety Standards in Managed Pressure Drilling

Safety standards in managed pressure drilling (MPD) are critical for ensuring the well-being of personnel, the integrity of equipment, and the protection of the environment. These standards are designed to mitigate risks associated with high-pressure drilling environments and the complexities of real-time operations. Regulatory bodies and industry organizations have established comprehensive guidelines that address various aspects of safety, including equipment specifications, operational procedures, and training requirements for personnel. Adherence to these standards is not only essential for compliance but also for the promotion of a culture of safety within the drilling industry.

One of the key components of safety standards in MPD is the requirement for advanced equipment that meets strict performance criteria. Equipment used in MPD operations, such as blowout preventers, pressure control systems, and automated monitoring tools, must be tested and certified to withstand the unique challenges posed by fluctuating pressures and dynamic drilling conditions. Regular maintenance and inspections are mandated to ensure that this equipment functions properly and can respond effectively in emergency situations. The implementation of robust equipment standards helps to minimize the potential for equipment failures, which could lead to catastrophic incidents.

Training and competency of personnel are also central to safety standards in MPD. Operators must undergo rigorous training programs that cover both theoretical knowledge and practical skills relevant to managed pressure drilling techniques. This training not only includes the operation of specialized equipment but also encompasses emergency response protocols and safety management systems. Continuous professional development is encouraged to keep personnel updated on the latest safety practices, technological advancements, and regulatory changes. A well-trained workforce is instrumental in identifying potential hazards and implementing effective mitigation strategies.

Moreover, real-time data analytics play a pivotal role in enhancing safety standards within MPD operations. By leveraging advanced data analytics tools, operators can monitor drilling parameters continuously and detect anomalies that may indicate potential safety risks. The integration of predictive analytics enables proactive decision-making, allowing teams to anticipate issues before they escalate into serious problems. This data-driven approach not only improves operational safety but also contributes to more efficient drilling processes, ultimately reducing the risk of incidents caused by human error or equipment malfunction.

Lastly, the commitment to safety standards in managed pressure drilling extends beyond compliance; it reflects a broader responsibility towards environmental stewardship and sustainability. The industry is increasingly focused on minimizing its environmental footprint through responsible drilling practices. Implementing stringent safety measures helps to prevent spills, blowouts, and other environmental incidents that could have far-reaching consequences. By prioritizing safety, the oil and gas sector not only protects its workforce and assets but also reinforces its commitment to sustainable practices that benefit both current and future generations.

Compliance Challenges and Solutions

Compliance challenges in managed pressure drilling (MPD) are multifaceted, stemming from the need to adhere to stringent regulations, safety standards, and environmental guidelines. The complexity of drilling operations, alongside the rapid evolution of technology, creates a landscape where compliance can often be difficult to navigate. Regulatory bodies impose strict requirements to ensure safety and environmental protection, and the dynamic nature of drilling operations can lead to scenarios where compliance becomes a moving target. Researchers and professionals in the field must stay informed about the latest regulations and best practices to mitigate risks and enhance operational efficiency.

One significant compliance challenge arises from the integration of real-time data analytics into MPD systems. While real-time analytics can significantly enhance decision-making and operational control, they also raise concerns regarding data integrity, privacy, and security. Ensuring that the data collected and analyzed adheres to regulatory requirements is critical. This involves implementing robust data governance frameworks that not only protect sensitive information but also ensure that the data used for compliance reporting is accurate and verifiable. Addressing these challenges requires collaboration between technology providers, drilling operators, and regulatory bodies to establish clear standards for data management.

Another challenge pertains to environmental compliance, particularly in the context of sustainability initiatives. As the oil and gas industry faces increasing scrutiny over its environmental impact, MPD operations must incorporate sustainable practices that align with regulatory expectations. This includes minimizing emissions, managing waste effectively, and ensuring that drilling activities do not adversely affect local ecosystems. Professionals in the field must adopt advanced technologies that facilitate real-time monitoring of environmental indicators, enabling proactive measures to mitigate potential impacts. By integrating environmental considerations into compliance frameworks, companies can enhance their reputation and operational viability.

Training and education also play a critical role in addressing compliance challenges in MPD. As regulations evolve, continuous education for personnel involved in drilling operations is essential to ensure that they are equipped with the knowledge of current compliance requirements and best practices. Developing comprehensive training programs that focus on regulatory standards, safety protocols, and environmental stewardship can empower teams to make informed decisions that prioritize compliance. By fostering a culture of compliance through education, organizations can reduce the risk of violations and enhance overall operational safety.

Finally, collaboration among stakeholders is vital to overcoming compliance challenges in managed pressure drilling. This includes engaging with regulatory agencies, industry associations, and technology providers to share insights and develop best practices. Establishing forums for dialogue can facilitate the exchange of information regarding compliance strategies, lessons learned, and innovative solutions. By working together, stakeholders can create a more cohesive approach to compliance in MPD, ensuring that safety, environmental, and regulatory standards are met while advancing operational efficiency and effectiveness.

Chapter 8: Future Trends in Managed Pressure Drilling

Technological Advancements on the Horizon

The landscape of managed pressure drilling (MPD) is undergoing a transformation driven by rapid technological advancements. Emerging technologies are enhancing the efficiency, safety, and environmental sustainability of drilling operations. Innovations such as machine learning algorithms and artificial intelligence are being integrated into real-time data analytics systems, enabling operators to make informed decisions based on predictive modeling and historical data. These advancements allow for more precise control over drilling parameters, reducing the risk of wellbore instability and improving the overall success rate of drilling programs.

Another significant advancement is the development of advanced sensor technologies that facilitate real-time monitoring of drilling conditions. These sensors provide critical data regarding pressure, temperature, and flow rates, allowing for immediate adjustments to be made in response to changing conditions. The increased accuracy and reliability of these sensors contribute to enhanced operational transparency and facilitate better communication among team members. This interconnected approach not only streamlines workflows but also fosters a culture of safety and accountability within drilling operations.

The integration of cloud computing and big data analytics is also poised to revolutionize managed pressure drilling. By harnessing vast amounts of data generated during drilling activities, operators can identify patterns and trends, thereby optimizing drilling performance. Cloud technology enables remote access to data, facilitating collaboration among geographically dispersed teams. Furthermore, the scalability of cloud solutions allows for the integration of additional data sources, providing a more holistic view of drilling operations and enabling more strategic decision-making.

Environmental considerations are becoming increasingly central to drilling operations, with technological advancements aimed at minimizing ecological impact. Innovations in drilling fluids and materials are being developed to reduce the environmental footprint of drilling activities. Biodegradable and non-toxic drilling fluids are gaining traction, offering safer alternatives that align with sustainability goals. Additionally, advancements in waste management technologies are improving the handling and disposal of drilling waste, ensuring compliance with regulatory standards while minimizing harm to the environment.

As these technological advancements continue to evolve, their implications for regulatory compliance and safety standards in managed pressure drilling cannot be overstated. Enhanced monitoring capabilities not only aid in adhering to industry regulations but also contribute to proactive risk management strategies. The integration of automated safety systems is expected to reduce human error, further increasing the safety of drilling operations. As the industry embraces these innovations, ongoing research and collaboration among scientists, engineers, and regulators will be essential to fully realize the potential of these technologies in advancing managed pressure drilling practices.

Evolving Industry Standards

Evolving industry standards in managed pressure drilling (MPD) are crucial as the oil and gas sector faces increasing demands for efficiency, safety, and sustainability. These standards are shaped by advancements in technology, regulatory requirements, and the need for improved environmental practices. As researchers and professionals delve into real-time analytics, it becomes imperative to understand how these evolving standards are influencing the methodologies and technologies employed in MPD. The integration of real-time data analytics plays a significant role in adhering to these standards, providing insights that drive operational excellence and decision-making.

The adoption of advanced technologies in managed pressure drilling has led to the development of new industry standards that prioritize safety and efficiency. Innovations such as automated drilling systems and enhanced sensor technologies have transformed traditional practices. These advancements require a recalibration of existing standards to incorporate the capabilities of modern equipment. For instance, real-time monitoring systems enable operators to make informed decisions based on live data, thus minimizing risks associated with pressure fluctuations and drilling inefficiencies. As such, the evolution of industry standards must keep pace with technological advancements to ensure that safety and efficacy are continuously enhanced.

Regulatory compliance is another critical aspect of evolving industry standards in MPD. Regulatory bodies are increasingly focusing on the environmental impacts of drilling activities, necessitating stringent compliance measures. Standards are being refined to address concerns related to emissions, waste management, and resource conservation. Researchers and professionals must stay abreast of these regulations to ensure that their practices align with the latest requirements. This alignment not only mitigates legal risks but also enhances the industry's reputation by demonstrating a commitment to environmental stewardship and sustainability.

Moreover, the sharing of best practices through case studies on successful managed pressure drilling projects has become a vital component of evolving industry standards. These case studies provide valuable insights into the implementation of new technologies and methodologies, illustrating their effectiveness in real-world applications. By documenting successes and challenges, industry stakeholders can refine their approaches and contribute to a collective understanding of what constitutes best practices in MPD. This continuous feedback loop fosters a culture of innovation and improvement, driving the evolution of standards that reflect the dynamic nature of the industry.

In conclusion, the landscape of managed pressure drilling is undergoing significant transformation as industry standards evolve in response to technological advancements, regulatory pressures, and environmental considerations. For researchers, professionals, and scientists, engagement with these evolving standards is essential for fostering innovation and ensuring that practices remain relevant and effective. By leveraging real-time data analytics and embracing best practices from successful projects, the industry can navigate these changes effectively, leading to safer, more efficient, and sustainable drilling operations.

The Role of Research and Innovation

Research and innovation play a pivotal role in the advancement of managed pressure drilling (MPD), particularly as the oil and gas industry faces increasing demands for efficiency, safety, and environmental sustainability. The integration of real-time analytics into MPD systems has revolutionized how drilling operations are conducted, allowing for proactive decision-making based on continuous data streams. Researchers and professionals in this field are tasked with developing methodologies that not only enhance drilling performance but also minimize the ecological footprint of such operations. By investing in innovative technologies, stakeholders can achieve greater precision in pressure management, resulting in optimized drilling parameters and reduced non-productive time.

The development of advanced equipment and technologies is essential to meet the challenges posed by complex drilling environments. Research initiatives focused on improving sensors, data acquisition systems, and modeling techniques have led to significant enhancements in real-time monitoring capabilities. These innovations facilitate a deeper understanding of subsurface conditions and enable operators to respond swiftly to unanticipated events. The collaboration between academia and industry is crucial in this context, as it fosters the exchange of knowledge and accelerates the deployment of cutting-edge solutions. Researchers are encouraged to explore new materials and techniques that can withstand the extreme conditions often encountered in MPD operations.

Real-time data analytics has emerged as a cornerstone of effective managed pressure drilling. The ability to collect and analyze vast amounts of data instantaneously allows for immediate adjustments to drilling strategies, thereby enhancing safety and operational efficiency. Researchers are exploring machine learning and artificial intelligence algorithms to further refine data interpretation processes. By harnessing these technologies, practitioners can predict potential issues before they escalate, ensuring compliance with safety standards and regulatory requirements. This proactive approach not only protects personnel and assets but also contributes to the sustainability goals of oil and gas companies.

In addition to technical innovations, the environmental impact of managed pressure drilling is an area of active research. As the industry shifts toward more sustainable practices, understanding the ecological implications of drilling operations becomes increasingly important. Studies that examine the lifecycle impacts of drilling technologies, including emissions and waste management, are essential for developing strategies that align with environmental regulations and corporate social responsibility initiatives. Researchers play a critical role in assessing the trade-offs between operational efficiency and environmental stewardship, helping to shape policies that promote sustainable development within the sector.

Finally, case studies on successful managed pressure drilling projects serve as valuable resources for practitioners looking to implement innovative techniques. By documenting real-world applications of research findings, these case studies provide insights into best practices and lessons learned. They highlight the importance of continuous improvement and the adoption of new technologies in achieving operational excellence. For researchers, analyzing these projects can reveal gaps in current knowledge and inspire further investigation into emerging challenges. Ultimately, the synergy between research, innovation, and practical application is vital for advancing managed pressure drilling in a rapidly evolving industry landscape.

Chapter 9: Conclusion

Recap of Key Insights

The integration of real-time analytics within managed pressure drilling (MPD) has revolutionized the oil and gas industry, offering unprecedented insights that enhance operational efficiency and safety. A thorough understanding of these insights is essential for researchers and professionals aiming to master advanced techniques in MPD. The key findings illustrate not only the operational benefits of implementing real-time data analytics but also the associated advancements in equipment and technologies that facilitate this integration. Effective data utilization leads to improved decision-making processes, ultimately resulting in reduced drilling risks and enhanced performance.

A significant insight is the impact of real-time data on optimizing drilling parameters. By continuously monitoring variables such as pressure, flow rate, and temperature, operators can make informed adjustments that enhance the drilling process. This dynamic capability allows for immediate responses to unexpected changes in subsurface conditions, thereby minimizing non-productive time and improving drilling efficiency. The ability to analyze data as it is generated facilitates a more proactive approach to managing drilling operations, which is crucial in today's fast-paced and often volatile oil and gas market.

Moreover, the intersection of real-time analytics and environmental sustainability in MPD cannot be overlooked. The insights gained from real-time data not only aid in operational efficiency but also play a critical role in minimizing environmental impact. Advanced analytics can identify potential hazards and optimize resource usage, ensuring compliance with environmental regulations. This dual focus on efficiency and sustainability positions real-time analytics as a cornerstone of responsible drilling practices, aligning industry goals with broader environmental objectives.

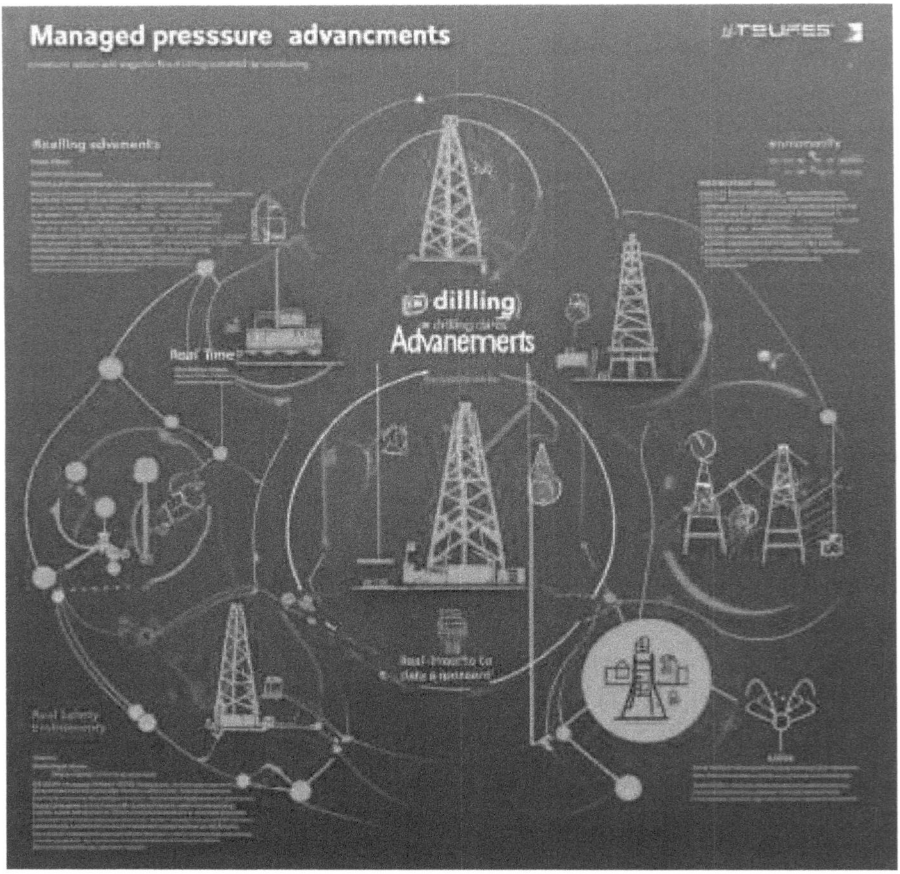

Case studies showcasing successful MPD projects highlight the practical applications of real-time analytics in addressing specific challenges within the drilling process. These examples provide empirical evidence of the benefits derived from advanced data analysis, demonstrating how organizations have achieved significant cost savings and improved safety records. By examining these case studies, researchers and professionals can glean best practices and innovative strategies that can be applied in their own projects, fostering a culture of continuous improvement and knowledge sharing within the industry.

In conclusion, the insights gathered from the integration of real-time analytics into managed pressure drilling underscore the transformative potential of this approach. As the industry evolves, the ongoing development of advanced technologies and analytical tools will further enhance the capabilities of MPD, ensuring that safety, efficiency, and sustainability remain at the forefront of drilling operations. For researchers and professionals, staying informed about these key insights is vital for driving innovation and maintaining competitive advantages in an increasingly complex landscape.

Future Directions for Research and Practice

The future directions for research and practice in real-time analytics for managed pressure drilling (MPD) present numerous opportunities to enhance efficiency and safety in the oil and gas industry. As the demand for more precise drilling techniques grows, researchers must focus on developing advanced algorithms and machine learning models that can process vast amounts of data generated in real-time. This involves integrating data from various sources, such as drilling sensors, geological surveys, and environmental monitoring systems, to create a comprehensive understanding of drilling conditions. The evolution of data analytics tools will enable professionals to make informed decisions on-the-fly, thereby optimizing drilling parameters and reducing non-productive time.

Another critical area for future research is the development of predictive analytics that can anticipate drilling challenges before they occur. By harnessing historical data and real-time measurements, researchers can establish predictive models that identify potential issues such as wellbore instability or formation pressure changes. This proactive approach not only enhances operational efficiency but also significantly mitigates risks associated with drilling operations. Collaboration between academia and industry will be essential in developing these models, ensuring they are grounded in real-world applications and tested under various drilling conditions.

The environmental impact of managed pressure drilling remains a significant concern. Future research should focus on minimizing the ecological footprint of drilling operations while maximizing resource recovery. This includes investigating sustainable practices, such as the use of biodegradable drilling fluids and the implementation of closed-loop systems that reduce waste. Additionally, researchers should explore the integration of real-time environmental monitoring systems that can detect and respond to potential hazards, ensuring compliance with regulatory standards and promoting environmental stewardship within the industry.

Technological advancements will also play a pivotal role in shaping the future of MPD. The development of advanced equipment, such as automated drilling systems and smart sensors, can significantly enhance the accuracy and reliability of data collection. Future research should prioritize the design of these technologies to ensure they are resilient in harsh drilling environments. Moreover, the role of the Internet of Things (IoT) in connecting these devices for seamless data exchange should be examined, paving the way for more integrated drilling operations that leverage real-time insights for improved performance.

Lastly, comprehensive case studies on successful MPD projects will be invaluable for guiding future research and practice. By analyzing these cases, researchers can identify best practices and lessons learned that can inform new projects and strategies. Documentation of the challenges faced and the solutions implemented in these projects will create a repository of knowledge that can be utilized by professionals in the field. Encouraging a culture of knowledge sharing and continuous improvement will be essential in driving the adoption of innovative techniques and technologies in managed pressure drilling, ultimately leading to safer, more efficient, and environmentally responsible operations.

Final Thoughts on Managed Pressure Drilling

Managed Pressure Drilling (MPD) has emerged as a transformative technique within the oil and gas sector, distinguished by its capability to enhance operational efficiency while mitigating risks associated with traditional drilling methods. As we conclude our exploration of MPD, it is essential to reflect on the significant advancements and insights gained through the integration of real-time analytics. The application of sophisticated data analytics not only aids in optimizing drilling parameters but also plays a critical role in decision-making processes, ultimately leading to improved safety and performance outcomes.

The advancements in technology surrounding MPD have facilitated a paradigm shift in how drilling operations are conducted. By leveraging real-time data, operators can make informed adjustments to drilling parameters, thereby maintaining wellbore stability and minimizing the likelihood of incidents such as kicks or blowouts. The continuous feedback loop created by real-time monitoring systems enables professionals to respond promptly to unexpected conditions, highlighting the importance of integrating advanced equipment with robust analytics capabilities. This integration is not merely beneficial; it has become essential for the successful execution of complex drilling projects.

Moreover, the environmental implications of MPD cannot be understated. The ability to control pressure more effectively results in reduced surface pollution and minimizes the risk of hydrocarbon leakage into surrounding ecosystems. As researchers and practitioners in the field focus increasingly on sustainability, it is imperative to emphasize the role of MPD in promoting environmentally responsible drilling practices. The data-driven approach allows for a more precise assessment of environmental impact, ensuring compliance with regulatory standards while also fostering community trust in oil and gas operations.

Case studies demonstrating successful MPD projects further reinforce the method's effectiveness in a variety of challenging environments. These examples provide valuable insights into best practices, showcasing how tailored strategies and real-time analytics have led to enhanced drilling performance and reduced costs. By examining these real-world applications, professionals can glean lessons that inform future endeavors, ensuring that the benefits of MPD are maximized across diverse geological and operational contexts.

In conclusion, the future of managed pressure drilling is poised for continued evolution, driven by ongoing research and technological innovation. As the industry confronts increasing regulatory pressures and societal expectations surrounding environmental stewardship, the significance of MPD will only grow. For researchers, professionals, and academicians, the challenge lies in harnessing the full potential of real-time analytics to not only advance drilling efficiency but also to contribute to a sustainable and responsible energy future. The journey towards mastering managed pressure drilling will require collaboration, knowledge-sharing, and a commitment to continuous improvement in practices and technologies.

www.ingramcontent.com/pod-product-compliance
Lightning Source LLC
LaVergne TN
LVHW050923210926
848482LV00012B/517
* 9 7 9 8 8 9 5 8 7 7 5 0 0 *